Sonatas for Violin and Basso Continuo Opus 3

Recent Researches in Music

A-R Editions publishes seven series of critical editions, spanning the history of Western music, American music, and oral traditions.

Recent Researches in the Music of the Middle Ages and Early Renaissance
Charles M. Atkinson, general editor

Recent Researches in the Music of the Renaissance
James Haar, general editor

Recent Researches in the Music of the Baroque Era
Christoph Wolff, general editor

Recent Researches in the Music of the Classical Era
Eugene K. Wolf, general editor

Recent Researches in the Music of the Nineteenth and Early Twentieth Centuries
Rufus Hallmark, general editor

Recent Researches in American Music
John M. Graziano, general editor

Recent Researches in the Oral Traditions of Music
Philip V. Bohlman, general editor

Each edition in *Recent Researches* is devoted to works by a single composer or to a single genre. The content is chosen for its high quality and historical importance, and each edition includes a substantial introduction and critical report. The music is engraved according to the highest standards of production using the proprietary software MusE, owned by Music | Notes.™

For information on establishing a standing order to any of our series, or for editorial guidelines on submitting proposals, please contact:

A-R Editions, Inc.
801 Deming Way
Madison, Wisconsin 53717

800 736-0070 (U.S. book orders)
608 836-9000 (phone)
608 831-8200 (fax)
http://www.areditions.com

Giovanni Battista Somis

Sonatas for Violin and Basso Continuo Opus 3

Edited by Glenn Burdette

A-R Editions, Inc.
Madison

Performance parts are available from the publisher.

A-R Editions, Inc., Madison, Wisconsin 53717

A-R Editions is pleased to support scholars and performers in their use of *Recent Researches* material for study or performance. Subscribers to any of the *Recent Researches* series, as well as patrons of subscribing institutions, are invited to apply for information about our "Copyright Sharing Policy."

Printed in the United States of America

ISBN 0-89579-422-5
ISSN 0484-0828

♾ The paper used in this publication meets the minimum requirements of the American National Standard for Information Sciences—Permanence of Paper for Printed Library Materials, ANSI Z39.48-1984.

Contents

Acknowledgments

Were it in my power, I would thank all of the scholars cited in these pages whose work has made mine possible, but the greater part of them are enjoying far finer rewards. I was glad to inform Boris Schwarz about Somis's opus 3 sonatas in the spring of 1983; his kind response was an encouragement. I owe much to Professor Neal Zaslaw, through whom I first encountered these sonatas in a musicology seminar devoted to eighteenth-century ornamentations of Corelli's opus 5. I also owe a debt of gratitude to Professor Donald Foster for many helpful suggestions relating to this endeavor during the course of my doctoral work. I also wish to thank the Syndics of Cambridge University Library for their gracious permission to publish this edition based on Cambridge Add. MS 7059.

Introduction

Apparently unpublished and possibly unperformed in their own time, the twelve solo violin sonatas that comprise opus 3 of Giovanni Battista Somis (1686–1763) deserve to be performed and known in ours. But because of their relative inaccessibility until now, the existence of these sonatas has not been reflected in standard reference works or known to performers.

The unique extant copy of these sonatas, part of a manuscript compilation of several Italian compositions,[1] was acquired by F. T. Arnold in the course of researching his monumental *Art of Accompaniment from a Thorough-Bass* (1931) and was then bequeathed to Cambridge University. The miscellaneous character of this compilation, now designated Cambridge Add. MS 7059, and bound together with several engraved sonatas, is captured in the *Report of the Library Syndicate [of Cambridge University] for the Year 1943–1944:*

> Somis (G. B.), Tartini, Vivaldi, &c. [Volume containing 18th century MS copies of works by Somis, Tartini, Vivaldi, Scaccia, Duncan, Fanfani, Pepusch; with Alberti's 12 Sonatas for violin and bass, c. 1720, engraved.]

Altogether, Somis published two collections of six trio sonatas (opuses 5 and 8), *Ideali trattenimenti da camera* for two violins or flutes and descant viols (opus 7), a set of cello sonatas, and four printed collections of twelve chamber sonatas each for solo violin and basso continuo: [opus 1], Amsterdam, ca. 1717, reissued 1725;[2] opus 2, Turin, 1723; opus 4, Paris, 1726; and opus 6, Paris, 1734. The sonatas of opus 3 fit precisely into this chronology. Another collection of sonatas that found their way into Holland in a manuscript copy should perhaps be considered an additional publication.[3]

The Composer

Somis was born into a family already renowned for its service in Savoy-Piedmont. His father, Lorenzo Francesco (1662–1736), served as orchestral violinist to Madama Reale, the duke of Savoy's mother, and as a household official of the prince of Carignano, the duke's cousin. Giovanni Battista's brother Lorenzo (1688–1775) also became an accomplished violinist, and their sister Christina Antonia (1704–85) became a remarkable singer.

Giovanni Battista joined the ducal cappella in 1696 as violinist, and in 1703 Duke Vittorio Amedeo II sent him to Rome "per habilitarsi nella musica,"[4] as a pupil of Corelli. Somis remained there until 1706 or 1707. Beginning in 1709, he added service as "aiutante di camera" of the prince of Carignano to his duties at the ducal court. The prince's frequent residence in Paris and involvement with that city's social and musical scene was to have a profound influence upon Somis later. In 1711 Somis showed up among the musicians invited to a religious festival in Novara,[5] and in 1713 he visited Sicily in (now King) Vittorio Amedeo's retinue.[6] His first collection of violin sonatas, the unnumbered opus 1 (ca. 1717), was dedicated to Madama Reale, and opus 2 (1723), to Vittorio Amedeo. Promising students began to seek out the *primo violino;* among the most important of these were Jean-Marie Leclair, who encountered Somis in Turin in 1722 and 1726, and Jean-Pierre Guignon, a Piedmontese who was a rival to Leclair at the French court. Gaspard Fritz became a fixture in Geneva, as did Felice Giardini in London. It is likely that Gaetano Pugnani studied with Somis, and it is also possible that Louis-Gabriel Guillemain did.

Though written or copied in Turin, opus 3 was directed at Paris.[7] The dedicatee, Louis-Henri, duke of Bourbon, a grandson of Louis XIV, was serving as prime minister for Louis XV. Louis-Henri's mistress, the Marquise de Priè, wife of the former ambassador to Turin, had founded a series of concerts of Italian music "qui pensait dégoûter les Parisiens de Molière et de Lully," and at which Christina Somis sang.

Piedmontese musicians had long been in contact with French music and society. From about 1715 they were active in Paris, where the prince resided in the Hôtel de Soissons. Though the prince employed Lorenzo Somis *fils* as violinist in the carnival season of 1715,[8] we have no secure evidence of Giovanni Battista's performing in Paris during this period. On the other hand, Somis must have enjoyed some renown: Le Cène reissued Somis's opus 1 in 1725, and in the following year his first French publication (opus 4) appeared. And the vast sums amassed by the prince from 1717 to 1720 in the frenzied speculation known as the "Mississippi Bubble" could have enabled considerable expansion in the musical and social activities of the Carignanos throughout the 1720s. The year 1725 also marks the inception of the Concerts Spirituels, where from the start Guignon was popular, and it became an almost obligatory stop for any Piedmontese performer. Guignon led the way for Pugnani and Giardini; Somis's relatives Carlo Chiabrano (Chabran), Giuseppe Canavasso, and Pietro Miroglio (as composer); and Somis himself.[9]

In 1730 the prince of Carignano began a series of concerts at the Hôtel de Soissons. Until 1741 these Sunday concerts were a celebrated fixture in Parisian life,[10] and may have been the proximate reason for Somis's first performance opportunities in France in the spring of 1733.[11] The dedication of his opus 5 trio sonatas (1733) to the prince may reflect this connection. But it is certain that Somis gave concerts at the Concerts Spirituels during this visit—concerts which earned him "applause and admiration" and the enthusiastic praise of the *Mercure de France.* Performing various sonatas and concertos, presumably of his own composition, Somis was imposing by the "grace and elegance of his execution, the power of his bowing, and his touching expressivity."[12]

Rejoining the cappella upon his return to Turin, Somis was rewarded with the position of first solo violinist and private musician to the new king, Carlo Emanuele III.[13] We find Somis in the court theater in 1736, performing as *primo violino,* and serving in the capacity of *direttore musicale e di scena* in 1737–57.[14] He must have continued to be in demand as a teacher, though the majority of his published compositions were then behind him. A royal wedding in 1750 brought Veracini to Turin as guest violinist[15] and prompted Somis's opus 7, *Ideali trattenimenti da camera a due violini, o due flauti traversieri o due pardessus de viola,* after a long hiatus in publication. Somis continued in service to the prince of Carignano and the king of Sardinia (duke of Savoy) until his death in 1763.

The Music

Boris Schwarz wrote that "Somis transplanted Corelli's Roman tradition to the north and shaped it in his own image," observing that Somis "reduced the number of movements to three, usually an introductory Adagio followed by two Allegro movements . . . [and] expanded the first Allegro into a three-section form—a statement, a digression and an abbreviated reprise in the principal key, comparable to an incipient sonata form in the Classical sense."[16] Keeping the size of the outer movements modest, although performers skilled at improvisation could have expanded them somewhat by ornamenting the adagios and varying the dances, Somis lengthened the middle movements. Most commonly this lengthening entails the use of sequence, an occasional echo or *petite reprise,* or one or more sections of passagework that resemble variations (see the allegro movement of opus 3, Sonata VIII). The resulting movements are usually bipartite, i.e., the two halves are roughly equal, though a significant minority are of the "tripartite" type.[17] In most cases, the thematic recapitulation occurs before the tonal recapitulation; seldom do they coincide. The most usual type of fast movement for Somis was binary with parallel openings[18] in the two halves and some degree of "rounding" at the end. The nature of the "rounding" of his binary movements varies. In some cases, development and recapitulation sections are separated by a clear cadence, other times not; occasionally the second half of the movement is a single sequential passage that reaches the tonic two or three measures before the movement ends. Third movements in opuses 1–3 are typically rounded binary and are dance movements. As such they are rhythmically plain and rather brief, at least in their written form, due largely to their restrained use of sequence. In many cases they serve only as unornamented themes for sets of variations, indicated by the terms "Grazioso" or "Gratioso."[19]

One must add, however, that Somis's melodic style is equally innovative, especially in adagio movements. The overall result shows that it was something quite different from Corelli's Roman tradition that Somis was taking or sending north—a newer style, or *style galant,* representing a transitional period between baroque *Fortspinnung* and classical periodicity. In Somis's slow movements one sees a tendency for the melody to form short, motivic, irregular phrases (usually clearly separated from each other by rests and large leaps), particularly at the beginnings of movements. Almost invariably the opening phrase and its "answer" are followed by a third of greater length, producing the effect of completing what the preceding phrases only started. This pattern established, Somis then proceeds with his own brand of gradual variation: soon the soloist's range widens, incorporating successively wider leaps, and the speed increases through the use of smaller note values. But at some point in the movement, attention is drawn away from the opening phrase or motive; the movement then continues freely.

The Parisian collections of opuses 4 and 6 exhibit a greater variety of forms throughout, including several movements in rondo form. The melodies are adorned with *agréments* in typically French fashion, with some instances of reverse dotting,[20] *Trommelbass,* and some dynamic indications. There is little counterpoint, but passages of what Veracini called "lazy composition," that is, in which the solo and bass parts are conducted in parallel thirds or tenths for several measures, are found, as are chains of suspensions and long passages of pure figuration, some in triplet rhythms. But most significant is the extension of the sonata principle to slow movements: they are occasionally rounded binary in form, and are often in a contrasting key. Thus more and more sonatas are arranged fast-slow-fast. Another innovative feature of opus 4 is that Somis builds movements of individual sonatas on similar themes.

Notes on Performance

Like most other collections of its time, Somis's opus 3 bears the misleading title "Sonate da camera a violino solo, e violoncello, o' cembalo," implying that either the cello or harpsichord should be employed, but not both together. Most writers consider this a stock phrase of no special significance to performance practice, though Michelangelo Abbado takes the phrase seriously, and remarks that

> there is no doubt that, in most cases, performance with the violin and the harpsichord would be preferable in order to obtain the complete harmony demanded by the composer in his basso continuo figures. It should be pointed out, nevertheless, that here and there the bass part presents passages of writing that are decidedly idiomatic to the violoncello and are little adapted to the harpsichord; in such cases one can follow the version for the two stringed instruments. In any case, I believe that the rather widespread practice nowadays of the performance with all three instruments is inadvisable.[21]

The composer's intentions regarding improvised ornamentation in these sonatas are likewise unknown. The ornaments notated in Cambridge Add. MS 7059 are few, and reserved for the soloist. They are limited to the trill, the mordent, the appoggiatura, the slide, and, in one instance (in the adagio of Sonata IV), the *bariolage,* that is, "repeated notes played alternately on two strings, one stopped, one open,"[22] indicated by a long wavy line. The trills are notated clearly, with upper auxiliaries frequently written out, as they are in this edition. The precise placement of the continuo figures in the manuscript reveals that appoggiaturas should be played long.

It should not be forgotten that free ornamentation must also have played an important role in endearing the ostensibly spartan melodies of opus 3 or other of Somis's earlier works to the jaded audiences of the 1733 Concerts Spirituels. It was, at any rate, a prominent feature of Somis's playing, as shown in contemporary remarks concerning his playing.[23] Somis's study with Corelli, after all, had taken place during the years between the publication of the latter's opus 5 in Rome by Gasparo Pietra Santa (1700) and the edition of opus 5, with graces "comme il les joue," published by Etienne Roger in Amsterdam (1710).

With regard to dynamics, the cautious performer may well choose to abide by what has been called "baroque terrace dynamics,"[24] allowing the occasional appearances of echo effects and *petites reprises* to introduce a certain degree of dynamic contrast. A comparison of the first few measures of Somis's opus 3, Sonata II, with those opening Leclair's opus 2, Sonata VII, in a similar type of movement with marked dynamic indications, shows to what extent this might be done.[25]

Notes

1. These are discussed in Glenn Burdette, "The Violin Sonatas of Giovanni Battista Somis (1686–1763), Including an Edition of Opus 3" (Ph.D. diss., University of Cincinnati, 1993), 142–51. Opus 3 comprises the first twenty-five folios of the manuscript.

2. The information given for the unnumbered opus 1 refers to a reissue by Le Cène using Roger's plates, both printings bearing the catalog no. 456. See *Répertoire international des sources musicales: Einzeldrücke vor 1800,* ed. Karlheinz Schlager et al. (Kassel: Bärenreiter, 1971–86) 8:115. Le Cène provided a new title page, substituting his own name for that of his mother-in-law, and styling Somis "maître de musique."

3. In 1759 the auction catalog of Nicolas Selhof's library and instrument collection advertised among its "musique en manuscrit" a collection of "J. B. [*sic*] Somis, Sonate a Violino o Flauto Traverso Solo col Basso Continuo, partitura," but, of course, there is no reason this could not be a manuscript copy of previously printed sonatas. *Catalogue d'une très belle bibliothèque de livres, curieux & rares, en toutes sortes de facultez & langues . . . [et] livres de musique. . . .*(The Hague: Veuve d'Adrien Moetjens, 1759; reprint, Amsterdam: Frits Knuf, 1973, entitled *Catalogue of the Music Library, Instruments and Other Property of Nicolas Selhof, Sold in The Hague, 1759*), 208, catalog no. 2136.

4. Alberto Basso, "Notizie biografiche sulle famiglie Somis e Somis di Chiavrie," in Giovanni Battista Somis, *Sonate da camera opera II per violino e violoncello o cembalo,* ed. Michelangelo Abbado (Milan: Edizioni Suvini Zerboni, 1977), p. xvi, col. 1, citing the Archivio di Stato di Torino, Sezioni Riunite, Tesoreria Generale della Real Casa, Articolo 217, reg. 1703, art. 449. In Rome, Somis would have met Cardinal Pietro Ottoboni, dedicatee of opus 4 (1726), and an illustrious company of performers and composers: the cardinal's *maestro di cappella* Giuseppe Ottavio Pitoni, alongside Geminiani, Gasparini, Castrucci, Carbonelli, Anet, probably Alessandro and Domenico Scarlatti, and possibly Locatelli and Handel. The belief that Giovanni Battista Somis studied under Vivaldi was probably precipitated by an inaccurate statement of Burney that "Lorenzo [*sic*] Somis, maestro di cappella to the King of Sardinia, was regarded in Italy as of Corelli's school, a little modernised, after the model of Vivaldi." Charles Burney, *A General History of Music from the Earliest Ages to the Present Period,* ed. Frank Mercer (New York: Harcourt, Brace and Company, 1935), 2:446.

5. Guglielmo Barblan, "La musica strumentale e cameristica a Milano nel '700," in *Storia di Milano* ([Milan]: Fondazione Treccani degli Alfieri per la Storia di Milano, 1962), 16:620–21. Among the sixteen violinists was Angelo Maria Scaccia, a minor composer represented in Cambridge Add. MS 7059.

6. *The New Grove Dictionary of Music and Musicians,* s.v. "Somis, Giovanni Battista," by Boris Schwarz.

7. Elsewhere I have speculated on possible dates of composition for opus 3, other than the "1725" specified on its title page. By considering opuses 1–3 as a stylistic group, one can surmise a possible range of ca. 1717 (the date of opus 1) through 1725. Burdette, "The Violin Sonatas," 123–25.

8. See Basso, "Notizie biografiche," p. xiii, col. 2, n. 52.

9. Among the non-Piedmontese students who appeared at the Concerts Spirituels may be included Leclair, the Swiss violinist Gaspard Fritz, and the Frenchman Guillemain (as composer).

10. Georges Cucuel, *La Pouplinière et la musique de chambre au XVIIIe siècle* (Paris: Librairie Fischbacher, 1913; reprint, Geneva: Slatkine Reprints, 1971), 25.

11. I have not found information to concur with the statement that "the April 1725 Mercure described a concert at the Palais des Tuileries in which the same violinists [Somis and Guignon] were accompanied by only a bassoon and a bass viol" (Julie Ann Sadie, *The Bass Viol in French Baroque Chamber Music* [Ann Arbor: UMI Research Press, 1980], 28). The violinist who played with Guignon on that occasion was Anet.

12. *Mercure de France,* April 1733. In addition to the published sets, of course, these works could conceivably have been any of the solo violin sonatas of opus 6 (1734).

13. Bianca Becherini, "Un musicista italiano del XVIII secolo: Giovan Battista Somis," *Chigiana: Rassegna annuale di studi musicologici* 16 (1959): 8, although other sources state that Somis was named solo violinist in 1707 upon his return from Rome.

14. Basso, "Notizie biografiche," p. xix, col. 1.

15. Giulio Roberti, "La musica in Italia nel secolo XVIII secondo le impressioni di viaggiatori stranieri," *Rivista musicale italiana* 7 (1900): 727. Cited in Mary Gray White, "The Life of Francesco Maria Veracini," *Music and Letters* 53 (1972): 34.

16. *The New Grove,* s.v. "Somis, Giovanni Battista," by Boris Schwarz.

17. This is a distinction made by Charles Rosen in *Sonata Forms,* rev. ed. (New York: W. W. Norton, 1988), 16–27.

18. In only rare cases, as in the second of the cello sonatas or in the third movement of opus 6, no. 3, he dispenses with parallel openings.

19. Regarding this usage of the terms, see Burdette, "The Violin Sonatas," 122–23.

20. John Walter Hill writes that "the earliest appearance of written-out reverse dotting that I know of in the solo/bass sonata literature is in Locatelli's Opus II of 1732," i.e., only two years before Somis's opus 6 collection. "The Anti-Galant Attitude of F. M. Veracini," in *Studies in Musicology in Honor of Otto E. Albrecht,* ed. John Walter Hill (Kassel: Bärenreiter, 1980), 178.

21. Michelangelo Abbado, prefatory notes to Giovanni Battista Somis, *Sonate da camera opera II,* p. xxxv, col. 2: "non v'è dubbio che, nella maggior parte dei casi, sia preferibile l'esecuzione col violino e il clavicembalo, per ottenere la compiutezza armonica richiesta dal compositore con la numerazione del basso continuo. Si deve rilevare, tuttavia, che qua e là la parte del basso presenta passaggi di scrittura decisamente violoncellistica, poco adatti al cembalo: in tali casi ci si potrà attenere alla versione per i due archi. E' invece sconsigliabile, secondo me, la pratica oggi alquanto diffusa della esecuzione con i tre strumenti."

22. David Boyden, *The History of Violin Playing from Its Origins to 1761 and Its Relationship to the Violin and Violin Music* (London: Oxford University Press, 1965), 266–67.

23. Hubert Le Blanc, *Défense de la basse de viole contre les entreprises du violin et les prétentions du violoncel* (Amsterdam: Mortier, 1740; reprint, Geneva: Minkoff Reprints, 1975), 83–84, 96–98.

24. Although certain remarks by Charles de Brosses may support the use of gradual *crescendi* and *diminuendi.* See *Lettres familières écrites d'Italie en 1739 et 1740,* 3rd ed., ed. R. Colomb (Paris: Librarie Académique, P. Didier et Compagnie, 1869), 2:332–33.

25. See Jean-Marie Leclair, *Sonatas for Violin and Basso Continuo, Opus 2,* ed. Robert E. Preston, Recent Researches in the Music of the Baroque Era, vol. 58 (Madison: A-R Editions, 1988).

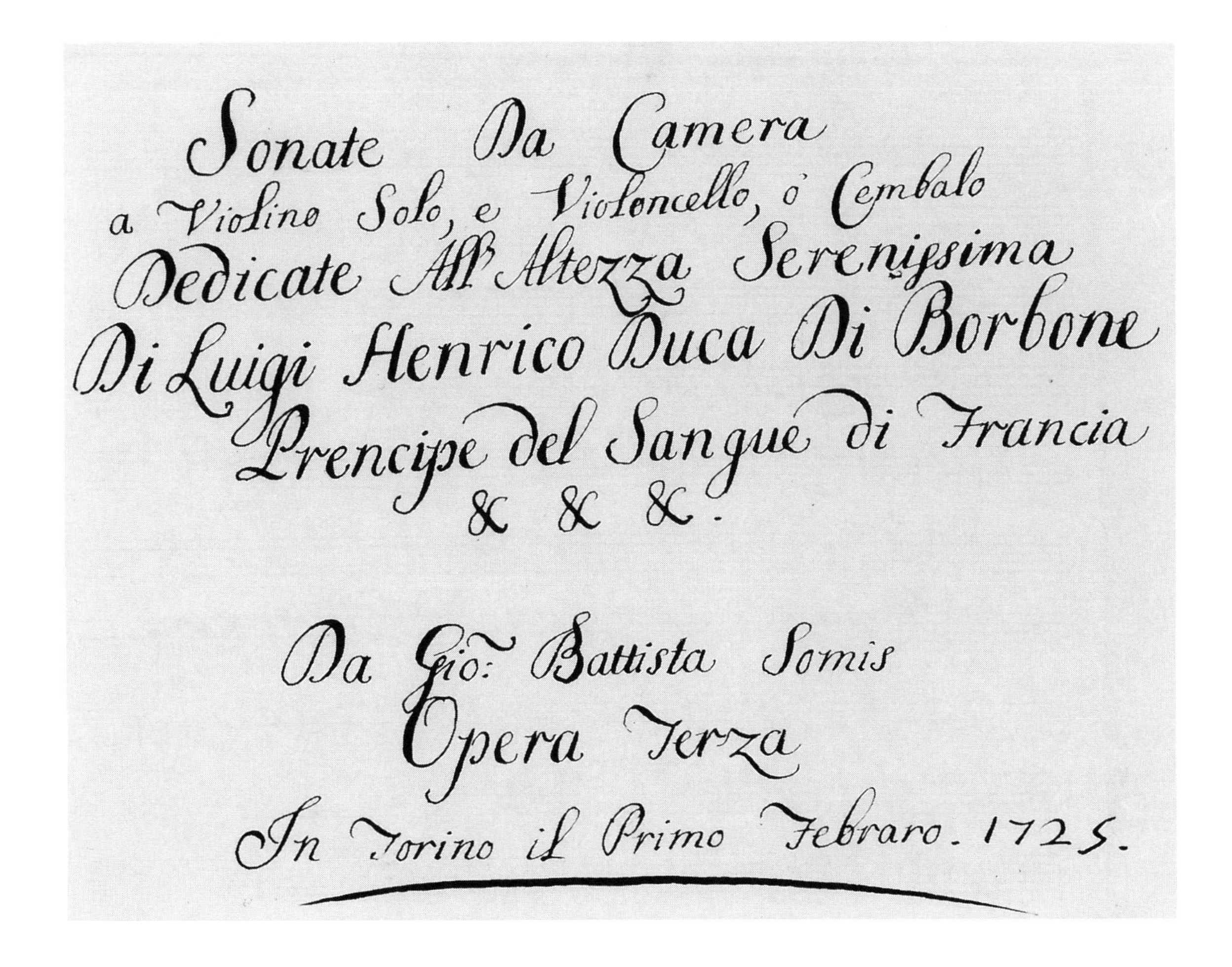

Sonate Da Camera
a Violino Solo, e Violoncello, o' Cembalo
Dedicate All' Altezza Serenissima
Di Luigi Henrico Duca Di Borbone
Prencipe del Sangue di Francia
&c &c &c.

Da Gio: Battista Somis
Opera Terza
In Torino il Primo Febraro. 1725.

Plate 1. Giovanni Battista Somis, *Sonate da camera a violino solo, e violoncello, o' cembalo . . . opera terza* (Turin, 1725), title page (from Cambridge Add. MS 7059). Reproduced by permission of the Syndics of Cambridge University Library.

Plate 2. Giovanni Battista Somis, *Sonate da camera a violino solo, e violoncello, o' cembalo . . . opera terza* (Turin, 1725), first page of Sonata I (from Cambridge Add. MS 7059). Reproduced by permission of the Syndics of Cambridge University Library.

Sonata I

28
Allegro
5
9
13
17

21
25
29
33
37
41

45
48
52
56
60
64

Allegro

Sonata II

Allegro

Allegro assai

Sonata III

Allegro

27
31
6
5
34
#6
6
♮5
6
4
5
3
38
6
44
7
6
3
7
49
7
6
5
6
5
7
6
5
#6

Allegro

Sonata IV

Allegro

Gratioso

Sonata V

31
Allegro
6
9
13
17

20
23
28
33
38
41

Allegro assai

Sonata VI

Allegro

28
34
39
44
49
53

57
61
65
69
74
18

82
86
Gratioso
6
12
18

Sonata VII

Allegro

Allegro

15
20
25
30
35
40

Sonata VIII

Allegro

Gratioso

Sonata IX

Allegro

15
19
22
25
27
29

32
7 6
6
6 4
5 3
Allegro
6
6
6
6
6
6
5 4
3
4
6
6
6
6
6 5
5 4
♯3
8
6
6
6
6
6
12
6
6
6
6
16
6
6
6
6
6 5
5 4
3

Sonata X

Allegro

Allegro

Sonata XI

Allegro ma poco

Allegro
5
10
15
20
25

Sonata XII

Allegro

Allegro

Laus Deo

Critical Report

Editorial Practice

The figures for the basso continuo appear between the treble and bass staves in Cambridge Add. MS 7059, and that placement has been maintained in this edition. The manuscript makes no use of slashed or compound figures, and an accidental accompanying a figure may either precede or follow it; in this edition accidentals always precede numerals. A few corrections in figures are documented in the notes to individual movements, but apparent redundancies, as in measures 15–16 of the third movement of Sonata X, are retained.

The manuscript includes occasional figured bass accidentals that are placed within rather than above the bass staff. That these are not meant to inflect notes of the bass line but are meant for the figured bass is clear since they precede empty lines and spaces rather than noteheads. Several of these accidentals are shown in plate 2 of this edition; see, for example, measures 5, 11, and 15 of the basso continuo staff. This notational practice is not confined to a manuscript such as this one but also occurs in the 1734 edition of Somis's opus 6 sonatas published by Boivin and Le Clerc (see Sonata III, first movement, mm. 27–28). Reports of specific instances of such accidentals in opus 3 are provided in the critical commentary. The copyist also occasionally used accidentals on the staves in the manner of directs.

Key signatures have been altered when necessary to conform to modern practice, as detailed in the critical commentary. The accidental usage followed by the copyist of these sonatas presumes the validity of accidentals for the following note and any immediate (or nearly immediate) repetitions of it, whether or not a barline should intervene. In the present edition, the modern convention of an accidental remaining in force throughout a bar unless canceled has been adopted such that redundant accidentals within a bar have been tacitly removed; source accidentals remaining in force beyond a barline have been added as necessary in brackets, as have all other editorial accidentals. Cautionary accidentals have been added in parentheses. Where the figured bass makes use of flats to cancel sharps or sharps to cancel flats, the modern use of the natural sign is substituted with reports in the critical commentary.

Appoggiaturas, notated without slurs in the source, have had slurs added tacitly. The beaming in this edition is that of the manuscript, so that oddities of beaming often found in the bass part of the manuscript, such as occur when parallel passages are beamed successively in two-note and in four-note groupings, are retained. All text indications (tempo markings, "Laus Deo," etc.) are original, though their precise placement is editorial in some cases. But page turn indications, given variously as "Segue" or "Volti," have been omitted since they do not necessarily correspond to page turns in this edition.

Critical Commentary

In the following reports, notes (including ornamental notes) are numbered consecutively within a measure. Pitches are designated according to the system in which middle C = c′.

Source key signatures are altered as follows. Sonata I is notated in G dorian (one flat) but is transcribed with two flats. Sonata V is in D dorian (no flat) but is transcribed with one flat. Sonata X has two sharps but has been transcribed with three.

The intrastaff accidentals inflecting thirds of the basso continuo occur as follows. The flat sign appears in Sonata I, mvt. 1, mm. 5 (note 3), 15 (note 2), 18 (note 1), 31 (note 1), 32 (note 1); Sonata V, mvt. 2, mm. 48 (note 3), 51 (note 1), 55 (note 2); Sonata V, mvt. 3, mm. 9 (note 3) and 10 (note 8); Sonata VI, mvt. 2, m. 51 (note 1); Sonata VIII, mvt. 1, m. 27 (note 1); Sonata VIII, mvt 2, m. 29 (note 1); and Sonata XI, mvt. 1, m. 2 (note 5, third figure). The natural sign appears in Sonata I, mvt. 1, m. 11 (note 1) and Sonata XI, mvt. 2, m. 21 (note 2). The sharp sign appears in Sonata XII, mvt. 2, mm. 43 (note 4) and 45 (note 2).

Sonata I

ADAGIO

M. 16, basso continuo, note 1, figure is ♭5.

ALLEGRO (MVT. 2)

M. 49, violin, notes 5–8 are b♭′, c″, d″, b♭′. M. 58, basso continuo, the figure may be 5. M. 72, basso continuo, the figure 7 is uncertain.

ALLEGRO (MVT. 3)

M. 2, basso continuo, note 1, figure is ♭5.

Sonata II

LARGO

M. 1, basso continuo, note 6, figure is 6. M. 13, basso continuo, there appear to be figures, now illegible, over notes 8 and 9.

ALLEGRO

M. 4, basso continuo, a single figure 7 appears over note 2. M. 27, basso continuo, note 2, figure is 6.

Sonata III

ADAGIO

M. 8, basso continuo, note 1, figure is ♭7. M. 11, basso continuo, note 4, figure is ♭7.

ALLEGRO (MVT. 2)

M. 23, basso continuo, note 2, figure is ♭7.

ALLEGRO (MVT. 3)

M. 24, basso continuo, note 2, second figure is ♭7.

Sonata IV

ADAGIO

M. 19, basso continuo, note 3, figure is ♭7. M. 30, basso continuo, note 1, figure is $^{6}_{\flat 5}$.

Sonata V

ADAGIO

M. 14, violin, note 5 is lacking.

ALLEGRO

M. 51, basso continuo, note 1, figure is $^{7}_{\flat}$.

ALLEGRO ASSAI

M. 9, basso continuo, note 3, the flat sign may belong to the following figure.

Sonata VI

ALLEGRO

Mm. 2, 10, 30, 38, and 72, basso continuo, note 1 (appoggiatura) is lacking. Mm. 9, 29, 37, and 71, violin, note 1 (appoggiatura) is lacking. M. 44, basso continuo, note 5, figure is $^{6}_{\flat 5}$. M. 50, violin, notes 1–2 lack slur. M. 51, basso continuo, note 1, figure is $^{9}_{\flat}$.

Sonata VIII

ADAGIO

M. 18, basso continuo, note 2, figure is ♭7.

ALLEGRO

M. 19, basso continuo, note 2, figure is ♭7.

Sonata IX

ALLEGRO (MVT. 2)

M. 27, violin, the continuation throughout the measure of the sharp on note 4 is conjectural.

ALLEGRO (MVT. 3)

M. 15, basso continuo, note 2 is e.

Sonata X

ADAGIO—LARGO

M. 1, "Largo" indication appears between the staves. M. 15, basso continuo, note 7, figure is ♯5.

ALLEGRO (MVT. 3)

M. 28, basso continuo, note 1, figure is ♯.

Sonata XI

ADAGIO

M. 2, basso continuo, note 5, third figure is ♭. M. 12, violin, note 8 has ♯.

ALLEGRO MA POCO

M. 3, basso continuo, note 3, figure is $^{\sharp 4}_{2}$; note 5, figure is ♮7. M. 21, basso continuo, note 1, figure is ♭5.

Sonata XII

ADAGIO

M. 5, basso continuo, the figure ♯ appears beneath the figure 7 of note 1, not above note 3. M. 10, basso continuo, note 6, figure is $^{6}_{\flat 5}$.

ALLEGRO (MVT. 2)

M. 25, basso continuo, note 2, figure is ♯

ALLEGRO (MVT. 3)

M. 33, basso continuo, note 1, figure is ♯7.